EARTH OBSERVATIONS AND GLOBAL CHANGE

Why? Where Are We? What Next?

A Report of CSIS Space Initiatives

AUTHORS
Lyn Wigbels
G. Ryan Faith
Vincent Sabathier

July 2008

CSIS | CENTER FOR STRATEGIC & INTERNATIONAL STUDIES

About CSIS

In an era of ever-changing global opportunities and challenges, the Center for Strategic and International Studies (CSIS) provides strategic insights and practical policy solutions to decisionmakers. CSIS conducts research and analysis and develops policy initiatives that look into the future and anticipate change.

Founded by David M. Abshire and Admiral Arleigh Burke at the height of the Cold War, CSIS was dedicated to the simple but urgent goal of finding ways for America to survive as a nation and prosper as a people. Since 1962, CSIS has grown to become one of the world's preeminent public policy institutions.

Today, CSIS is a bipartisan, nonprofit organization headquartered in Washington, D.C. More than 220 full-time staff and a large network of affiliated scholars focus their expertise on defense and security; on the world's regions and the unique challenges inherent to them; and on the issues that know no boundary in an increasingly connected world.

Former U.S. senator Sam Nunn became chairman of the CSIS Board of Trustees in 1999, and John J. Hamre has led CSIS as its president and chief executive officer since 2000.

CSIS does not take specific policy positions; accordingly, all views expressed in this publication should be understood to be solely those of the author(s).

Photo credits: NASA/WilliamAnders—Apollo 8, cover: ©iStockphoto.com/George Jurasek, p. v; ©iStockphoto.com/Federico Montemurro, p. vi; NASA GSFC Scientific Visualization Studio, p. vii; Department of Defense/Staff Sgt. Oscar Sanchez, Iowa National Guard, p. xiii; Adam Sparkes, p. xiv; Department of Defense/Lance Cpl. Willard J. Lathrop, USMC, p. 3; Numerical Terradynamic Simulation Group, The University of Montana, p. 4; ©iStockphoto.com/Zoran Simin, p. 5; NASA, p. 6; NASA/Lawrence Ong, p. 6; David S. Roberts, p. 7; ©iStockphoto.com/Peter Cosgrove, p. 8; NASA GSFC/Craig Mayhew and Robert Simmon, p. 9; Erik Charlto, p. 17; NOAA, p. 18; NASA GSFC, p. 27; ©iStockphoto.com/Volker Kreinacke, p. 28.

Library of Congress Cataloging-in-Publication Data
CIP information available on request.
ISBN 978-0-89206-541-7

The CSIS Press
Center for Strategic and International Studies
1800 K Street, N.W., Washington, D.C. 20006
Tel: (202) 775-3119
Fax: (202) 775-3199
Web: www.csis.org

CONTENTS

PREFACE

Is it possible to predict or alleviate the impacts of natural and manmade disasters? From the recent earthquake in China to the cyclone in Myanmar to the rapid changes in our climate to the ongoing violence in Darfur, environmental and national security events are occurring around the globe. Can we learn to adapt to and mitigate the water shortages and droughts that, combined with crop failures and exacerbated by soaring energy prices and a growing demand for biofuels, have led to an unprecedented global food crisis? Will we be able to understand and take actions to minimize the impact of changing climate and associated weather events on the health of human populations—from addressing rising sea levels to the accelerated spread of disease? Will we be able to balance the need for a wider array of alternative energy sources with respect to surging energy prices, simultaneously managing the implementation of carbon emission agreements including carbon cap and trade agreements?

These questions, and many others, demonstrate the complex management challenges presented by global change.[1] In order for decisionmakers to address these management challenges, they must have reliable, continuous long-term data about our planet and environment. Earth observations—including sensors in space, on land, in the air, and at sea, as well as associated data management and dissemination systems, Earth system models, and decision support tools—provide the infrastructure to deliver the data needed to understand ongoing global changes. In the half century since the dawn of the space age, space-based technologies from communications satellites to the global position system (GPS)—which underpin the success of globalization in recent decades—have been instrumental in knitting our civilization more closely together. Similarly, we have started to rely on Earth observations as another global public good. Earth observations are critical in a number of areas including dramatic applications in managing the effect of disasters, monitoring global agricultural productivity, assessing natural conditions including the state of the Earth's fresh water supplies, and monitoring the indirect effects of global energy policies on Earth's climate. These are all part of the vast effort involved in understanding and managing the 20 percent to 80 percent of the U.S. economy (representing $2.75 trillion to $11 trillion in 2007) sensitive to weather in the short term, let alone the evolving risk profile associated with longer-term global change.

We have successfully developed and integrated space-based communications and navigation capabilities to bring us closer together. While we have made great strides in developing and using Earth observation capabilities, many challenges remain to provide equivalent accomplishments

1. For the purposes of this report, we use the phrase "global change" to encompass the entire range of environment-related change phenomena regardless of time scale—from climate change to changes in available natural resources to severe weather--that can be observed, monitored, or understood with the aid of Earth observation systems. Further, when terms such as predict, prevent, mitigate (or other similar words) are used, they are meant generally to cover the full range of activities that one performs based on actionable Earth observation data.

in the operational and sustained use of Earth observations for global security. While we have started to use Earth observations in predicting and responding to disasters, such as the Indonesian tsunami or Hurricane Katrina, we are far from secure in having an operational ability to systematically monitor, predict, mitigate, or understand in order to take the actions necessary to prevent the challenges caused by the ever-increasing pace of global change. If we are to understand and plan intelligently for global change, we must take every opportunity to build on our past successes and redress our existing shortcomings.

Today, there are a number of steps the United States must undertake to deliver on the potential of Earth observations. First, the United States has the opportunity to demonstrate strong leadership within the U.S. Earth observation community through coherent, integrated planning, budgeting, and management of an Earth observation system providing long-term, continuous data acquisition. Second, the United States must lead the world toward effective international cooperation on Earth observations and, consequently, global change. Like any other kind of strong international leadership, leadership in Earth observations enhances our national foreign policy capabilities from providing data to manage global resources to economic security enabled by Earth observation capacity building. Third, the United States must ensure that Earth observations meet the needs of all users and that the public and private sectors reinforce—not inhibit—each other to enable us to take advantage of the ingenuity and innovation that the private sector can offer.

Rather than learning to adapt to natural and manmade disasters, the changing climate, the global food crisis, and our growing appetite for energy, dealing only with the consequences after the fact, we need to start focusing our efforts on the Earth observation systems that will better connect humanity and its home, allowing us to prevent, predict, and mitigate the increasingly dramatic impacts of global change on a routine basis.

Bank damaged in 2005 by Hurricane Wilma—the most intense hurricane recorded in the Atlantic basin.

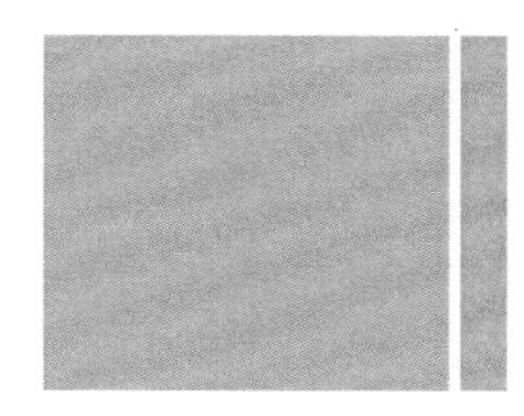

EXECUTIVE SUMMARY

The Value of Earth Observations and U.S. Challenges and Opportunities

The stresses on the Earth's systems are growing more severe at an ever-increasing pace, adding to the already significant economic variability arising from current challenges such as weather forecasting and resource management. The effects of these added pressures are already being felt and will have major implications for national security, the economy, natural resource management, and the security of water, food, and energy for decades to come. Today, U.S. public- (civil and national security) and private-sector users who want to understand global change or identify ways to predict, prevent, and mitigate its impacts are all intrinsically reliant on civil Earth observation systems (used in modeling, computation, and decision support tools) and data (collected from sensors on satellites, unpiloted aircraft, buoys, and other platforms). Earth observation products—including satellite weather information—provide, at a minimum, an additional $30 billion to the U.S. economy annually. In the future, Earth observation capabilities will be even more critical for governments and industry to monitor, understand, and adapt more quickly to global change and track and respond to consequences of past, present, and future policy choices. The national security community is increasingly concerned about the impacts of global change leading to instabilities and conflicts within, between, and among nations. This applies to stable as well as volatile regions.

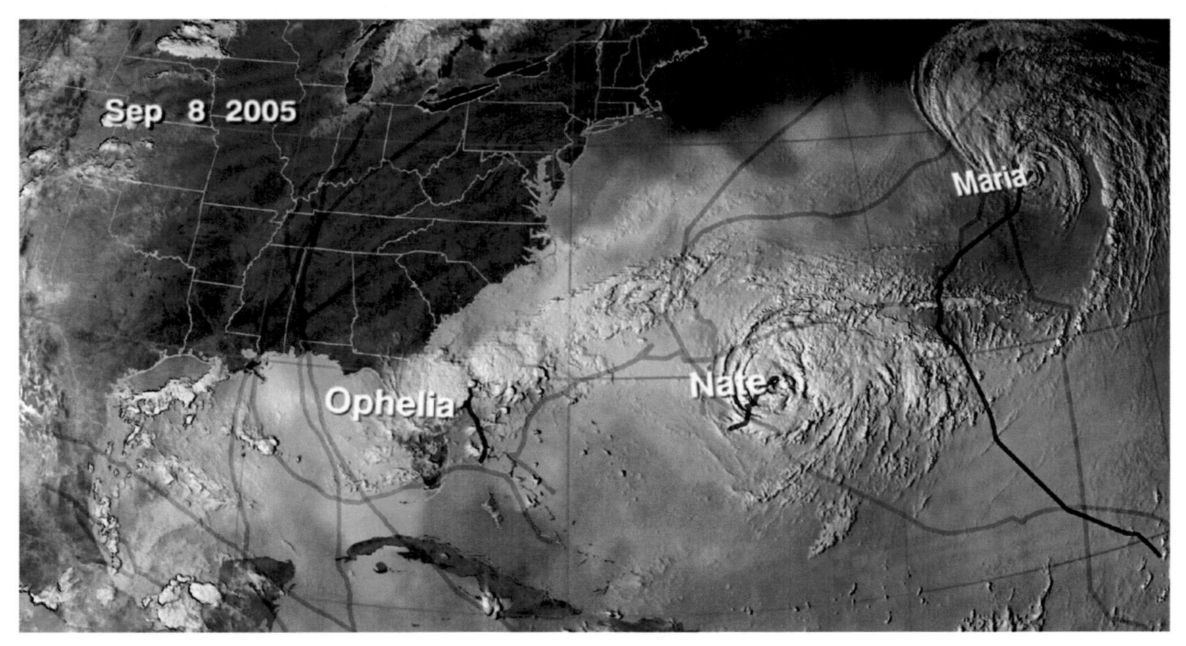

Satellite imagery showing three simultaneous hurricanes with sea surface temperatures, infrared cloud data, storm track data, and storm name labels.

The national security community is increasingly working with the Earth observation community to better understand these challenges.

Science communities have already determined a set of key observables that must be measured in order to effectively monitor the Earth system. The United States has a demonstrated Earth monitoring research capability and operates a highly effective national weather prediction system that has saved countless numbers of lives and billions of dollars. This aggressive research and development program has produced a number of proven sensors and ways of measuring essential variables, providing precise data that have yielded new scientific understanding and short-term forecasting improvements. However, due to structural and budgetary factors, these gains in obtaining new research data have not yet institutionalized plans for the continuous, complete, and comprehensive operational data sets needed to sustain monitoring and understanding of the longer-term—and perhaps much more important—climate changes that lie at the core of many current policy debates. The U.S. government has not yet established a commitment to comprehensive, long-term data acquisition for all essential variables. Data continuity will be critical for a full understanding of why, how, and how fast the Earth is changing. Similarly, there is not sufficient Earth observation capacity to operationally support many forms of Earth science and resource management. Furthermore, plans for a future comprehensive, coordinated, and sustainable U.S. Earth observation system to gather data for weather, climate, Earth science, and resource management continuously over longer time scales have not yet been established.

Having such a robust, comprehensive U.S. system is both the nation's responsibility to its citizens and the U.S. contribution to a Global Earth Observation System of Systems (GEOSS) and to the UN Framework on Climate Change Convention (UNFCCC), which the United States and the international community agree is needed during this time of significant global change. These responsibilities are not fully addressed by the allocation of resources for Earth observations in the United States.

The U.S. government is not currently organized to effectively lead, plan, fund, and implement an Earth observation program configured to provide the comprehensive sets of observable data essential for weather, climate, hazards, ecosystems, and related application areas. There is no single federal department, agency, or person in charge of addressing U.S national Earth observation activities as a whole. Instead, responsibilities and budgets are scattered among several federal agencies. This arrangement limits the support for a robust, holistic Earth observation program to meet national needs and the budget sufficient to implement it. This could become a critical issue with the potential implementation of the so-called cap and trade agreements for the management of carbon emissions. Cap and trade agreements will need both strong verification mechanisms and an understanding of how royalties from cap and trade programs will be managed. While the current U.S. Earth observation system can maintain, at least for the time being, a basic core capacity, there are a number of essential variables and capabilities that are not in place and for which there are no current plans. Initial estimates indicate that a critical threshold set of essential capabilities would require additional funding of more than $2.5 billion annually.

Recommendation 1. The United States should make a commitment to long-term, continuous data acquisition for all essential observations necessary to provide improved monitoring and prediction capabilities in order to sustain monitoring and evolve our understanding of the Earth system and how it is changing.

Recommendation 2. Building on the National Academy's decadal survey *Earth Science and Applications from Space*,[1] the United States should develop an overall plan for an integrated, comprehensive and sustained Earth observation system that (1) describes how these measurements can not only be acquired, archived, and distributed but also integrated as appropriate with Earth science models and decision support tools and (2) maps the goals and requirements for this system to and from the nine societal benefit areas identified by the U.S. Integrated Earth Observation System.[2] Shortfalls in the current plans should be addressed, and a vision for future generation Earth observation systems should be provided. Users from all sectors—public (federal, state and local governments), academia, and industry—should have the opportunity to provide input and participate in the definition and planning process.

Recommendation 3. The U.S. government should increase funding for Earth observations by doubling the budget from approximately $2.5 billion to $5 billion annually. This would enable expanding both space-based and in-situ observational capabilities to fully implement the National Academy's *Earth Science and Applications from Space* decadal survey recommendations and increase supercomputing, decision support tools, and modeling capabilities. This funding level should be reassessed following the development of an overall architecture for Earth observations in order to adequately fund an integrated, comprehensive, and sustained Earth observation system for weather, climate, hazards, Earth science, and resource management—consistent with the goals of GEOSS.

Recommendation 4. The United States should establish a governance structure under the supervision of a cabinet-level position that provides leadership both within the United States for Earth observations and for coordination and cooperation with other nations to promote and facilitate the planning, funding, and implementation of GEOSS. A formalized high-level interagency process should be established at the White House, led by a cabinet-level administration official responsible for U.S. Earth observation vision and goals, to establish agency roles and responsibilities for an integrated, comprehensive, and sustained Earth observation system and to develop the integrated budget to implement it. The president should provide an annual report on the state of the environment based on Earth observation measurements and on the status of and issues related to the development and operation of the comprehensive Earth observation system.

International Challenges and Opportunities

Global change presents a stewardship challenge requiring worldwide monitoring, not only to understand and predict future changes but also provide solutions to adapt globally to change and mitigate its impact. The environment is a global concern that cannot be meaningfully addressed or understood only at local levels; therefore, the United States cannot solve these problems alone but must work with other nations to solve them globally. Other nations are making commitments to and large investments in Earth observation systems that can complement U.S. Earth observation capabilities and contribute synergistically to a global system of systems. Engaging other nations in

1. Committee on Earth Science and Applications from Space, National Research Council, *Earth Science and Applications from Space: Urgent Needs and Opportunities to Serve the Nation* (Washington, DC: National Academies Press, 2005).

2. The nine areas are: (1) improve weather forecasting; (2) reduce loss of life and property from disasters; (3) protect and monitor our ocean resources; (4) understand, assess, predict, mitigate, and adapt to climate variability and change; (5) support sustainable agriculture and forestry and combat land degradation; (6) understand the effect of environmental factors on human health and well-being; (7) develop the capacity to make ecological forecasts; (8) protect and monitor water resources; and (9) monitor and manage energy resources.

developing and operating the global system of systems will promote the stewardship and solutions needed to address global change.

Earth observations can be an effective soft power tool in at least two particular respects: through the peaceful cooperation arising from coordination of civil Earth observation activities and through the use of Earth observation data as a development (including urban, agricultural, and natural resource planning) and disaster management aid. The United States considers data from civil Earth observation systems to be a global public good and has promoted an open data policy, although users in developing regions may lack the ability to maximize the use of this data without outside assistance. This open data policy, along with the level of U.S. investments in Earth observations and its participation in international fora, has characterized U.S. leadership in multilateral discussions on Earth observations.

The U.S.-led creation of the multinational Group on Earth Observations (GEO) has been a tremendous scientific, environmental, and foreign policy achievement. It has engaged governments at the ministerial level who have agreed on the value of Earth observations in obtaining concrete societal benefits. Through its strong leadership at GEO and in the Committee on Earth Observation Satellites (CEOS), the United States has made progress in engaging the international community in discussions on GEOSS. The United States has also led a growing consensus on making data freely available at low cost, which has prompted other nations (including Brazil, China, and Russia) to open up previously closed data sets.

However, much more needs to be done in order to implement GEOSS. Jason, the oceanography mission to monitor global ocean circulation, and NPOESS, the National Polar-orbiting Operational Environmental Satellite System, are examples of successful international engagement. There are further opportunities for the United States to be proactive in seeking partnerships on cooperative missions and developing interoperable systems. Further, with the exception of ocean monitoring, the United States has not built the cooperative relationships to transition new sensors and systems beyond what are essentially technology demonstration missions to long-term data acquisition and continuity missions. Additionally, current U.S. export control regulations are a significant structural impediment to and fundamental disincentive for U.S. collaboration with international partners, for international cooperation with the United States, and for the development of GEOSS. The International Traffic in Arms Regulations (ITAR) legislation has created real and perceived obstacles to engagement and cooperation. While ITAR was intended to cover critical, highly sensitive military technologies—a widely agreed on fundamental national right—in practice, the regulations are applied to a much wider array of other technologies. In addition, as individuals in the approval process are criminally liable equally for real and perceived mistakes, decisionmakers have a strong incentive to be excessively cautious. Furthermore, despite the written provisions of ITAR, the regulations are now being applied to data from space systems, not just the space systems themselves. These factors have led to the situation where ITAR has forced the international community to develop its own independent capabilities (for example, radar ocean altimetry and Lidar/IMU). Consequently, international companies now lead in several technologies, while U.S. firms are losing access to global markets and in some cases have lost the ability to produce such technologies altogether.

Recommendation 5. The United States should leverage its investment in Earth observations as a foreign policy tool, not only to promote global stewardship and develop global solutions but also to enhance U.S. soft power capabilities and improve its international image. The U.S. government should formally join the International Charter for Space and Major Disasters.

Recommendation 6. The United States should optimize international partnerships for the development and operation of current and planned Earth observation capabilities that leverage global synergies to minimize gaps and unnecessary overlaps while providing strategic redundancies.

Recommendation 7. The U.S. government should continue to support the multilateral GEO at a senior level to help actively promote the development of GEOSS and maintain the engagement of other governments at the ministerial level. Through its support of GEO, the United States should motivate the international community to make the investments necessary to build a coordinated, worldwide Earth observation system of systems. It should also continue to provide leadership and support to CEOS to define the space segment of GEOSS.

Recommendation 8. The multilateral GEO, with the support of CEOS, should build on the progress it has already made to champion and lead the process of developing a truly comprehensive, coordinated, worldwide Earth observation system of systems for all Earth observation needs. Building on the commitments already made by some GEO members, GEO should promote the adoption of an open data policy by all its members as a goal to be aspired to.

Recommendation 9. The State Department should continue to include the important role of Earth observations in general and the GEO in particular in responding to global change at the G-8 summit and in other high-level multilateral and bilateral fora to raise the visibility of Earth observations and global change at the highest levels of government.

Recommendation 10. The United States should revise its export control policies to promote dialogue among international governmental and industrial partners. The U.S. government should review the ITAR list to remove nonsensitive items related to Earth observations that present no threat to national security in order to promote dialogue among industrial partners, create a healthier climate for U.S. business, and achieve more international collaboration on Earth observations. Furthermore, ITAR should no longer inhibit the exchange of scientific data, in order to remove fundamental disincentives to the discussions that lead to cooperation.

Private-Sector Challenges and Opportunities

Earth observations provide key data critical to predicting and mitigating the impacts of global change. This has provided the private sector with an unprecedented opportunity to understand the impacts of global change on domestic and international markets, begin planning for investments and operations in a global economy greatly impacted by these changes, and reduce the uncertainty and operational risks the world's corporations and economies are likely to face. In addition to its political and policy benefits, the information from Earth observations is a public good of growing importance to economic and societal growth. User communities can now better manage and mitigate risks related to global change using Earth observations, as can clearly be seen in the examples of the stabilization of global commodities prices prompted by the work of the Department of Agriculture's Foreign Agricultural Service or the banking and insurance communities as demonstrated in the *Flood Insurance Reform and Modernization Act of 2007*. There is a growing recognition that as private-sector dependence on Earth observations grows, the level of support for such capabilities also needs to increase to sustain this continued growth, and the private sector will need to increase its support of these capabilities to ensure the continued and future availability of necessary information in a complete and timely fashion independent of uncertain global public investment.

There are many effective public-private partnerships involving the use and dissemination of data from government-purchased weather and land-imaging Earth observation platforms.

However, past U.S. government attempts to commercialize Earth observation capabilities (for example, moderate resolution land imaging) have not been successful. While attempts to commercialize Earth observations in other nations have succeeded to some extent, such as the French Spot Image, these cases have involved significant government support. There are many other mechanisms for effective collaboration among contributing segments of the community and emerging users, such as the recent National Oceanic and Atmospheric Administration (NOAA) partnership with Shell to place sensors on their offshore oil platforms, that will augment the national ocean monitoring capability.

Once the market matures and the economic value of Earth observations is more clearly established, there will be more opportunities for public/private partnerships and private sector initiatives. Other emergent opportunities, such as the possibility of food and water management, may be fertile areas for involvement of the private sector at an early stage of the development process. More effective understanding and integration of the entire Earth observation value chain will reduce the uncertainty that data and satellite industries face in making investments in Earth observation capabilities and applications and thus provide a broader base of support and capital investment for a more comprehensive global Earth observation capability meeting both the private and public sector needs.

Recommendation 11. The private sector should be an active participant in the development of the architecture for an integrated, comprehensive, and sustained Earth observation system to ensure its requirements are generated and incorporated in a holistic way and to enable insight into ways that a mature private-sector capability can evolve to address potential gaps. Recognizing U.S. government challenges in space systems acquisition management, the private sector should be called on to bring ingenuity and innovation to solutions for future system configurations, technologies, and delivery that focus on maximizing performance while reducing risks, shortening development time, and minimizing budgets.

Recommendation 12. The U.S. government should actively seek innovative public-private partnerships for developing and operating Earth observation systems that will capitalize on or promote the emergence of private-sector capabilities. In addition to the traditional commercial Earth observation players, the increasingly broad range of industries that are ever more reliant on Earth observations should be engaged to ascertain how they cooperate in their roles as consumers or suppliers of Earth observation products and data throughout the entire Earth observation system in meeting the overall objectives of Earth observation and global change public policy.

Recommendation 13. To promote commercialization, the U.S. government should build on its history of providing no-cost or low-cost data (weather and most recently, Landsat) to promote the further development of private-sector products (for example, AccuWeather). If applications prove to be as profitable as they have been in other technology fields (such as the global positioning system for navigation), they could support the business case for private development and operation of more focused and tailored private-sector Earth observation offerings.

Recommendation 14. The U.S. government should engage the finance community to discuss and address how its policies and activities promote or fail to promote private capital and equity market investment in support of goods derived from Earth observation. The government should take steps to remove some of the long-standing impediments to creating public-private partnerships, such as the inability of the federal government to make long-term commitments to use assets provided by the private sector.

Iowa Army National Guard soldiers assist local police with traffic and crowd control as search and rescue teams patrol flooded streets in search of stranded citizens in Cedar Rapids, Iowa, on June 12, 2008.

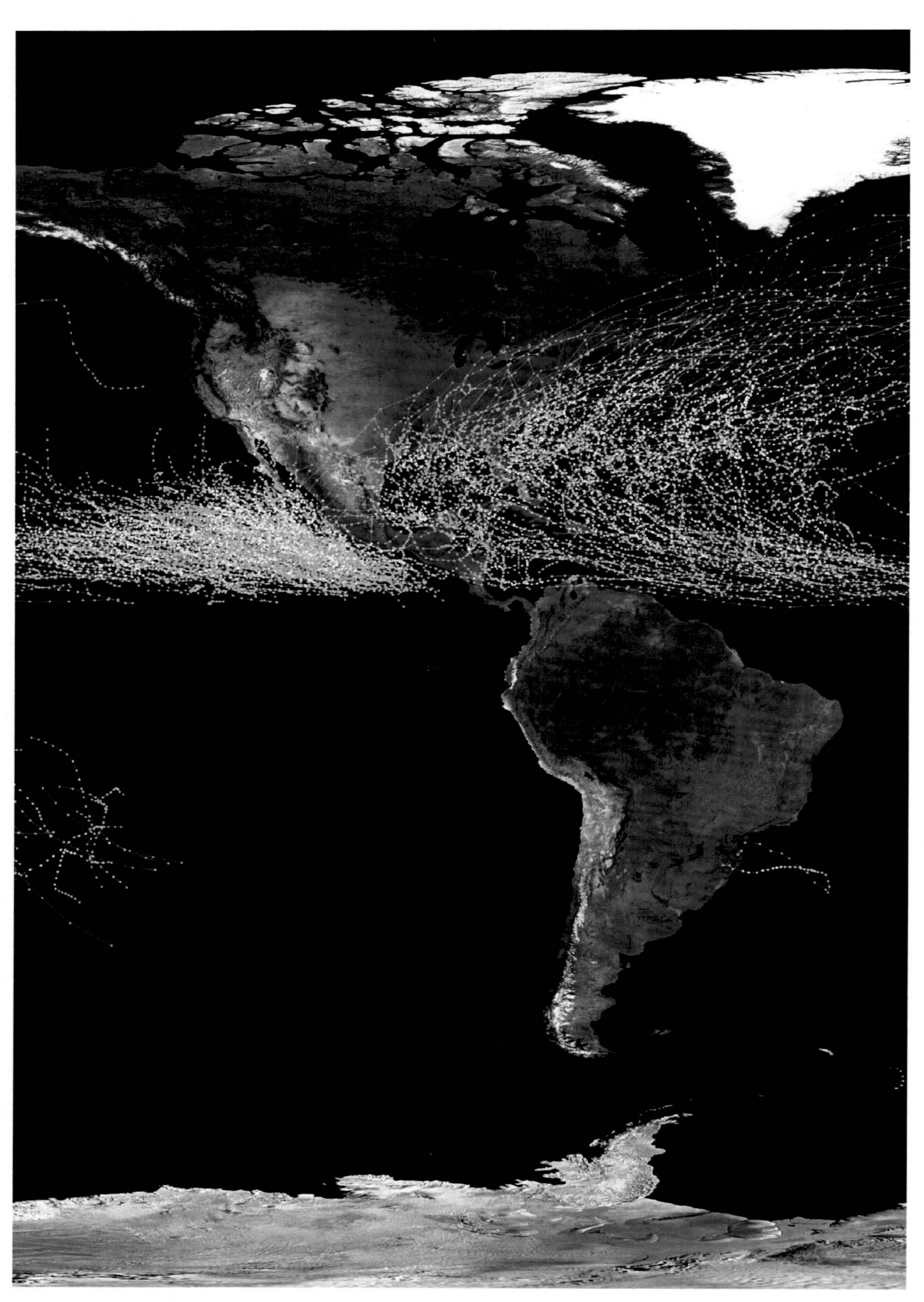

Map of tropical cyclone tracks and intensities from 1985 to 2005.

1 WHAT IS THE VALUE OF EARTH OBSERVATIONS

Background[1]

With the recognition that Earth observations are an important policy priority in the coming years, the Center for Strategic and International Studies (CSIS) initiated an Earth Observations and Global Change project in October 2007. The project began a dialogue among government, private-sector, and science communities to examine the role and value of Earth observations, assess the current state of the Earth observation system, and identify the gap between requirements and current and planned Earth observation capabilities. To this end, CSIS convened four working group meetings between October 2007 and May 2008. These meetings framed the issues associated with the current and next generation Earth observation system and explored national and international strategies to address these issues in order for the United States and the world to continue to benefit from this global public good. CSIS would like to thank the working group participants and contributors, listed in appendix A, for their contributions to this effort.[2]

Findings

Earth observations enable government and the private sector to understand and predict climatic and environmental events more effectively, prevent and mitigate hazards, prepare for disasters, and enhance security. In much the same way that management of our globally integrated service economy is inextricably tied to the ability of telecommunications and information technology to connect peoples and processes, our ability to function and prosper on Earth is inherently dependent on the use of Earth observation technology to enhance our ability to watch, observe, and understand our world. From the first cultivation of crops more than 10,000 years ago, the ability to observe, understand, and subsequently adapt to our world is a key requirement for civilization. It is in this sense that Earth Observations—the globalized, information-intensive observation of Earth--are necessary for the broad health and prosperity of all humanity and are inescapably a fundamental global public good.

Policy and management decisions are being made based on predictions and assessments derived from computer models using data from Earth observation systems. Support tools have been developed and are being used by multiple government agencies for decisionmaking in at least

1. For the purposes of this report, we use the phrase "global change" to encompass the entire range of environment-related change phenomena regardless of time scale—from climate change to changes in available natural resources to severe weather—that can be observed, monitored, or understood with the aid of Earth observation systems. Further, when terms such as predict, prevent, mitigate, or other similar words are used, they are meant generally to cover the full range of activities that one performs based on actionable Earth observation data.

2. In addition, we appreciate the invaluable assistance of our intern, Zoe Rose.

a dozen areas (agricultural efficiency, air quality, aviation, carbon management, coastal management, disaster management, ecological forecasting, energy management, homeland security, invasive species, public health, and water management). Fifty-five environmental data records identified by the Senior Users Advisory Group for the National Polar Orbiting Environmental Satellite System are being compiled in five mission areas: atmosphere, climate, land, ocean, and space environment. Importantly, the private sector is increasingly using this science for strategic planning and long-term decisionmaking.

Substantial scientific and economic research is being published on global change. Under the auspices of the Global Climate Observing System, the international scientific community identified 26 essential climate variables measurable by Earth observation satellites and other systems that they determined were necessary to understand global change. Observations and studies of the Earth are better documenting the changes that are putting greater and greater stresses on ecosystems around the world. Last year, the Intergovernmental Panel on Climate Change (IPCC) issued *Climate Change 2007: Synthesis Report*,[3] which completes its four-volume Fourth Assessment Report, summarizes the findings of the three working group reports, and provides a synthesis that specifically addresses the issues of concern to policymakers in the domain of climate change. The report concluded that "changes in the atmosphere, the oceans and glaciers and ice caps show unequivocally that the world is warming."[4] The IPCC Fourth Assessment Report documents new weather extremes including heat waves, new wind patterns, worsening drought in some regions and heavier precipitation in others, melting glaciers and Arctic ice, and rising global average sea levels. In several instances, the report recognized Earth observations as having a key role in providing higher confidence in ongoing scientific studies and new research findings. Observations are providing additional information for use in global models, and this increased information coupled with a larger number of more sophisticated models is providing a stronger quantitative basis for estimating likelihoods for future climate changes. Michael Jarraud, secretary general of the World Meteorological Organization (WMO), said that progress in observations and Earth observation measurements was key to improved climate research "and has considerably narrowed the uncertainties" of their previous report.[5] Importantly, the broad societal value of Earth observations extends beyond long-term data acquisition and continuity, across the information systems integrating in situ, space, communications assets and tools, and people.

Why Is the Private Sector Relying More on Earth Observations?

Concerns about global change, the vulnerability of societal infrastructure to sea-level rise and coastal inundations; changes in frequency, intensity, and probability of extreme events; and prolonged extreme conditions such as droughts have long been of paramount concern, often considered unavoidable risks and unexpected "Acts of God." However, in recent years, Earth observations have provided a key to better understanding of these events and in so doing have given the

3. Intergovermental Panel on Climate Change (IPCC), *Climate Change 2007: Synthesis Report,* Contribution of Working Groups I, II, and III to the Fourth Assessment Report of the Intergovernmental Panel on Climate Change (Geneva: IPCC, 2007), http://www.ipcc.ch/ipccreports/ar4-syr.htm.

4. UN Environment Program, "Evidence of Human-caused Global Warming 'Unequivocal,' Says IPCC," press release, February 2, 2007, http://www.unep.org/documents.multilingual/ default.asp?articleid=5506&documentid=499&l=en.

5. Ibid.

private sector an unprecedented opportunity to predict, manage, and mitigate such risks. Beyond more traditional business issues, greater private-sector interest in good corporate citizenship has increased attention to changes in air and water quality, loss of biodiversity, and spread of tropical diseases—other areas addressed by Earth observations. As a result, industries are increasingly using and relying on publicly financed Earth observation systems to help them make critical business decisions related to current activities and future plans.

U.S. Marines search through rubble for casualties following a massive mudslide that covered a village in the southern part of the island of Leyte in the Philippines.

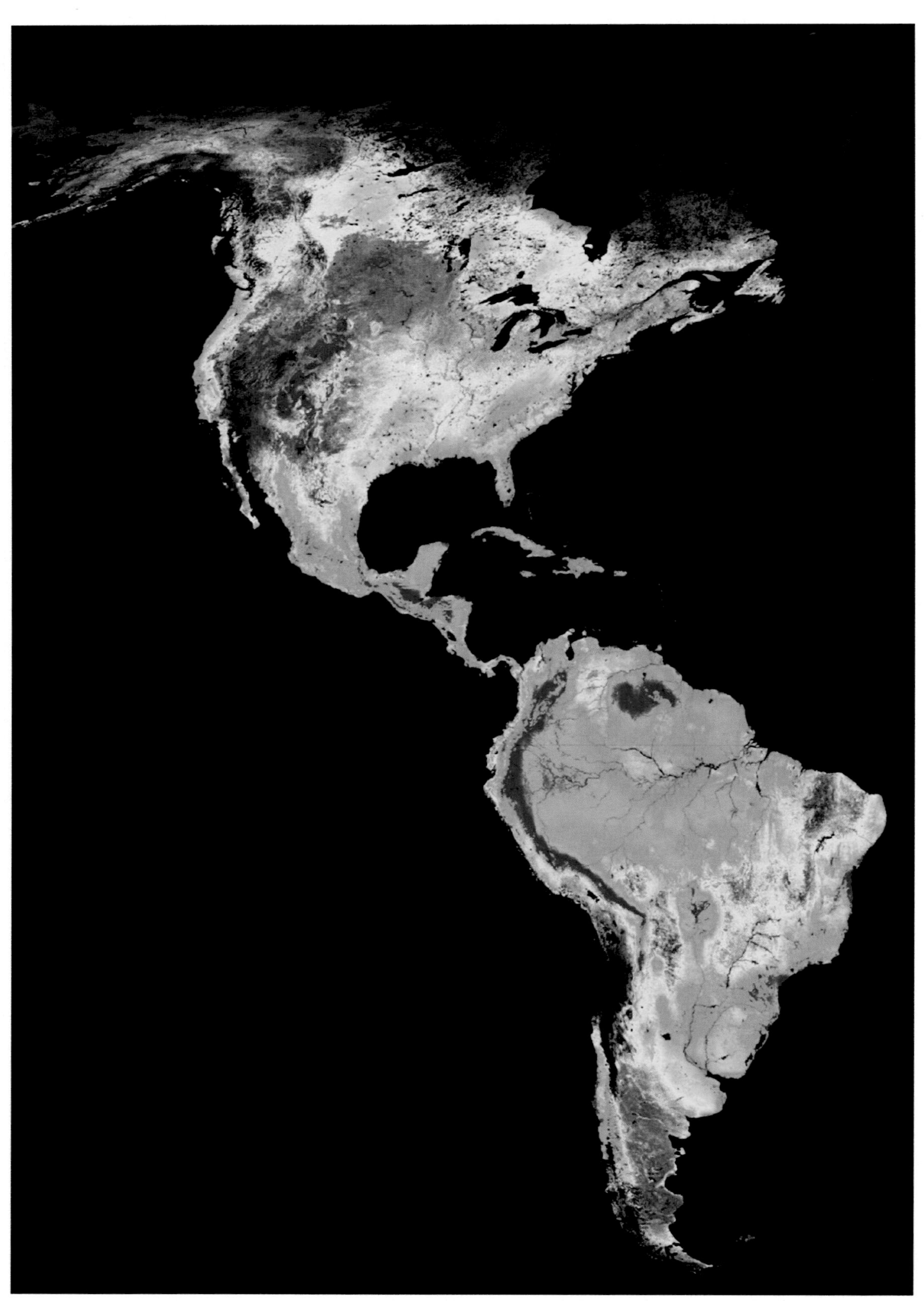

A measurement from the Moderate Resolution Imaging Spectroradiometer (MODIS) of Gross Primary Productivity (GPP), the sum total of the light energy that is converted to plant biomass.

Application: Food, Biofuels, and Agriculture

Earth observation systems have been used in some form or another by the agricultural sector for quite some time. Modern uses of Earth observations include measurement of product performance, crop yields, and the effectiveness of drought-resistant crop strains, seeds, and germplasms. Earth observations have been particularly valuable in understanding when, what, and why something happened with a particular crop. Earth observations are used to determine why a given customer's specific crop did not perform as expected and for the environmental stewardship of genetically modified crops used in the field. More recently, as growing demand for biofuels has precipitated a global food crisis, the need for agricultural companies and government planners to access Earth observation products to look at trends over cultivated areas around the world has become imperative.

Drought conditions, which have exacerbated an already precarious food supply situation created by the switch to the production of biofuels, have highlighted the need for effective land and water resource management tools.

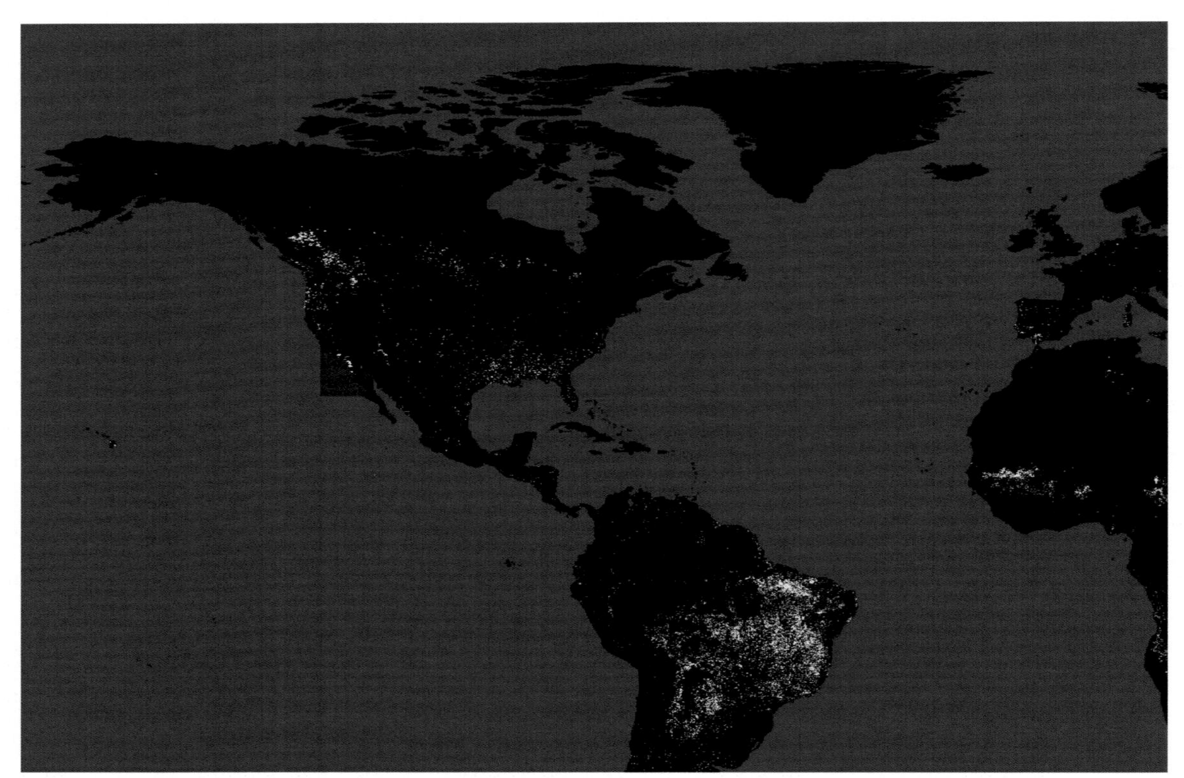
October 2007 map of fire intensity and locations.

October 23, 2007, image of the Harris Fire outside of San Diego, viewed with the Advanced Land Imager on the EO-1 satellite.

Application: Catastrophes, Disasters, and Insurance

One-third of total insurance risk is related to catastrophes. This risk has been growing at a rate of 10 percent to 15 percent per year and has doubled over the last five years (this alone could suggest that the need for investment in Earth observation capabilities has also doubled). In 2004 and 2005, financial losses directly attributable to natural and manmade catastrophes were $141 billion and $225 billion, respectively (in 2008 dollars). Most insurance companies use modeling companies to assess risk and have changed from looking retrospectively at historical weather data and simply extending past trends indefinitely into the future to acquiring Earth observation data and projecting future losses using sophisticated computer models. Recently, the global insurance industry has accepted the impact of global change on insurance rates not only for disasters. The survival of the insurance industry relies on its ability to make predictions of the economic impact of future events, and their financial interests are interdependent with understanding global change. Nevertheless, the insurance industry, particularly in the United States, has had a limited commitment to research and is wholly dependent on public sources of Earth observation data.

View of the Harris Fire burning down Mount Miguel outside of San Diego taken early on October 23, 2007, from 3 miles away.

Offshore oil platform, Ocean Warwick, blown ashore Dauphin Island, Alabama, by Hurricane Katrina.

Application: Economic Security, Carbon and Energy

Important questions for energy companies are not only how global change impacts are going to affect costs, supplies, and demand for different sources of energy but also how carbon control schemes, such as "cap and trade," will affect global economic activity. Energy companies need to know about the severity and frequency of storms for planning and demand forecasting. Earth observation data helps these companies take steps to protect oil and gas supplies, pipelines, offshore platforms, and terminals. For operations in Arctic areas, energy companies need to know whether permafrost will remain or change in order to properly design new facilities. Changes in air and water quality, loss of biodiversity, and spread of tropical diseases are also a concern in areas in which they operate, such as sub-Saharan Africa. Carbon mitigation is another area of increasing interest for the energy sector. Carbon capture and sequestration is a mitigation strategy that could help reduce greenhouse gas emissions, as well as increase oil field production. Earth observations can play a key role in determining whether carbon dioxide is escaping from storage. Further, as more and more energy and other companies become involved in carbon cap and trade agreements, the question of which system—global or national—will be used for verification is unresolved. Earth observations can also help with locating and monitoring arable land suitable for biofuel feedstock production; reforestation; data on oil spills and gas flaring; and the availability of oil, natural gas, geothermal, wind, hydroelectric, and other resources.

A map of the Earth's city lights was created with data from the Defense Meteorological Satellite Program (DMSP) Operational Linescan System (OLS).

Overview: Role of Earth Observations in Promoting Global Economic Growth

Recent studies have estimated that between 39 percent[6] and 16.2 percent[7] of the total U.S. gross domestic product (GDP) of $14.2 trillion is weather sensitive—some $5.5 to $2.3 trillion—an amount on the same scale as the entire 2009 U.S. federal budget. According to a CSIS analysis, Earth observation products—including satellite weather information—provide, at a minimum, an additional $30 billion to the U.S. economy annually.[8] This is across more than 10 sectors, including financial firms, insurance and reinsurance firms, the transportation industry, energy companies, and manufacturing. It should be noted that, much like Global Positioning System (GPS) and telecommunications, the amount of economic utility that can be derived from Earth observation systems grows as new applications and products are created. Given federal Earth observation expenditures of approximately $2.5 billion, the annual rate of return exceeds 10 to 1, comparable to the return on investment for satellite telecommunications. Scaling this up from the $13.13 trillion U.S. GDP to a global economy of $66 trillion, Earth observations may generate approximately $170 billion of annual economic activity. Hence, Earth observations appear to be a global public good. This analysis is considered to be quite conservative. It was based on a steady state environment, including only weather and climate effects and ignoring solid Earth and other types of Earth observations, and it did not take into account impacts from disasters. Functionally, the ultimate value of vastly increased capabilities in drought, climate, storms, blizzard, insect infestations and flood prediction; mineral, forest, fisheries, water, agriculture, grazing, fossil fuel, and wildlife management; or better natural disaster prediction, is in aggregate, very hard to estimate.

To maximize the benefits of Earth observations, the private sector has expressed a need for more accurate information at regional and local scales; higher spatial and temporal data resolutions; and a better understanding of the changing hazards, consequences, assets, and resiliencies associated with global change. They also point to the need for accuracy and timeliness of the information. More importantly, with the reliance on Earth observations growing, the private sector is concerned about how dependent it may become on a system that may be decaying—particularly one that is not strongly committed to long-term data acquisition and continuity. Industry requires continuity of these capabilities and more certainty about their availability in the future.

What Is the Role of Earth Observations in Maintaining National and International Security?

Concerns are increasing about the national security implications of tension and conflicts between and among nations such as large-scale migrations, resource scarcity, and disease pandemics caused by global change. Global change can increase tensions even in stable regions and is a threat multiplier in the most volatile regions.

---

6. John A. Dutton, "Opportunities and Priorities in a New Era for Weather and Climate Services," *Bulletin of the American Meteorological Society* 83, issue 9 (September 2002): 1303–1312.

7. Peter H. Larsen, "An Evaluation of the Sensitivity of U.S. Economic Sectors to Weather" (working paper, Social Science Research Network, May 2006), http://papers.ssrn.com/sol3/papers.cfm?abstract_id=900901.

8. John Hillery, "Measuring Earth Observation's Contribution to the U.S. Economy," September 2007, CSIS, http://66.28.70.47/component/option,com_csis_pubs/task,view/id,4532/type,0/.

This issue has been studied for more than a decade. In 1994, Berel Rodal recognized that environmental change has become a significant factor in international security.[9] In a 1996 study, Nicolini Strizzi and Robert Stranks considered the implications of environmental degradation related to fresh water supply on China.[10] In 1997, Peter Gizewski reviewed the effects of depletion and degradation of resources on societal processes, including population displacement leading to violence.[11] In 2004, Robert McLeman and Barry Smit examined climate change, migration, and security—highlighting migrations of 300,000 people in the southwestern United States in the 1930s, drought migrations in East Africa, and the displacement of 80,000 people by Hurricane Mitch in 1998.[12] And in spring 2007, the Strategic Studies Institute of the Army War College and the Triangle Institute for Security Studies held a colloquium entitled "Global Climate Change: National Security Implications."

Background: Think Tank Perspectives on Climate and Security

In April 2007, the CNA Corporation issued a report entitled *National Security and the Threat of Climate Change*, concluding that "projected climate change poses a serious threat to America's national security."[13] Nearly a dozen retired admirals and generals who comprise CNA's Military Advisory Board examined how climate change could affect U.S. national security over the next three to four decades. The report concluded that climate change, national security, and energy dependence are interrelated global challenges. The effects of climate change will exacerbate existing problems leading to massive migration, increased border tensions, and conflicts over essential resources, which can increase volatility in unstable regions and add to tensions in stable regions. These kinds of instabilities can contribute to the failure of states and the creation of breeding grounds for terrorism. The United States may be called upon to stabilize regions affected by climate change and suppress the outbreak of conflicts. The report recommended that the national security dimension of climate change be integrated into national security and defense strategies and that scientific agencies such as the National Oceanic and Atmospheric Administration (NOAA), National Aeronautics and Space Administration (NASA) and U.S. Geological Survey (USGS) also be brought into the processes for planning responses to global-change induced instability. Furthermore, the partnership between environmental scientists and the defense and intelligence communities, so vibrant during the 1990s, has provided critical security-relevant knowledge about climate change and should be revived.

In November 2007, the Center for a New American Security, in collaboration with CSIS, released *The Age of Consequences: The Foreign Policy and National Security Implications of Global*

9. Berel Rodal, "The Environment and Changing Concepts of Security," Commentary No. 47 (August 1994), Canadian Security Intelligence Service, Ottawa, http://www.csis.gc.ca/pblctns/cmmntr/cm47-eng.asp.

10. Nicolini Strizzi and Robert T. Stranks, "The Security Implications for China of Environmental Degradation," Commentary No. 67 (March 1996), Canadian Security Intelligence Service, Ottawa, http://www.csis.gc.ca/pblctns/cmmntr/cm67-eng.asp.

11. Peter Gizewski, "Environmental Scarcity and Conflict," Commentary No. 71 (Spring 1997), Canadian Security Intelligence Service, Ottawa, http://www.csis.gc.ca/pblctns/cmmntr/cm71-eng.asp.

12. Robert McLeman and Barry Smit, "Climate Change, Migration, and Security," Commentary No. 86 (March 2004), Canadian Security Intelligence Service, Ottawa, http://www.csis.gc.ca/pblctns/cmmntr/cm86-eng.asp.

13. Gordon R. Sullivan et al., *National Security and the Threat of Climate Change* (Alexandria, VA: CNA Corporation, April 2007), p. 6, http://securityandclimate.cna.org/report/ National%20Security%20and%20the%20Threat%20of%20Climate%20Change.pdf.

Climate Change.[14] The study examined three climate change scenarios (expected, severe, and catastrophic) and noted that the environmental effects caused by the "expected" climate change scenario are the least we ought to prepare for. National security implications include heightened internal and cross-border tensions caused by large-scale migrations, conflict sparked by resource scarcity, and the spread of disease. In case of "severe" climate change, the world is overwhelmed by the scale of change and challenges, such as widespread epidemics. The social consequences range from armed conflict between nations over resources to outright chaos. The "catastrophic" scenario finds strong intersections between two security threats—global climate change and international terrorism. Both threats are linked to energy, and solutions depend on America's energy economy. The report said that while the United States faces an ominous set of current and future national security and foreign policy challenges, global change may become the most daunting challenge on this list in the national security arena. The report notes that we lack rigorously tested data or reliable modeling to determine with any certainty the outcome of climate change, particularly at the scales most important to military planners.

Background: Congressional Perspectives on Climate and Security

National security implications of climate change have also become a growing concern in the U.S. Congress. The *Global Change Research and Data Management Act of 2007* (H.R. 906) would require the president to present Congress with a quadrennial assessment that analyzes "the vulnerability of different geographic regions of the world to global change." Both the Senate and House of Representatives included language in the *National Defense Authorization Act for Fiscal Year 2008*, signed by President George W. Bush on January 28, 2008, requiring the Pentagon to consider the effects of climate change on military capabilities, facilities, and missions and to develop the capabilities needed to reduce future impacts. The Senate and House of Representatives both included language in the *Intelligence Authorization Act for Fiscal Year 2008* (S. 1538 and H.R. 2082) that would require the director of national intelligence to submit to Congress within 270 days "a National Intelligence Estimate on the anticipated geopolitical effects of global climate change and the implications of such effects on national security of the U.S." The House of Representatives agreed to the conference report in December 2007; however, the legislation is still pending. The National Intelligence Council has already begun working with the U.S. Global Change Research Program and the Joint Global Research Institute on a study on the implications of climate change on national security, expected to be released later in 2008.

The House Subcommittee on Investigations and Oversight of the Committee on Science and Technology held a hearing, "The National Security Implications of Climate Change," on September 27, 2007, to hear from experts about the nature and magnitude of the threats that climate change may present to national security and to explore the ways in which such climate-related security threats can be predicted, forestalled, mitigated, or remedied. In his opening statement, Subcommittee Chairman Brad Miller said "this Committee fought for years the decision to eliminate sensors designed to collect climate-related data from the National Polar Orbiting Environmental Satellite System… Is the elimination of the sensors shortsighted on the basis of our national security needs?"

14. Kurt M. Campbell et al., *The Age of Consequences: The Foreign Policy and National Security Implications of Global Climate Change* (Washington, D.C.: Center for a New American Security, November 2007), http://www.csis.org/media/csis/pubs/071105_ageofconsequences.pdf.

What Is the Role of Earth Observations in Providing Global Public Goods?

Global change presents a stewardship challenge requiring worldwide monitoring, not only to understand and predict future changes but also provide global solutions to adapt to changes, prevent problems, and mitigate their impacts. The environment is a global concern that cannot be meaningfully addressed or understood at local levels only; therefore, the United States cannot solve these problems alone but must work with other nations to solve them globally. Other nations are making commitments to and large investments in Earth observation systems that can complement U.S. Earth observation capabilities and contribute synergistically to a global system of systems. Engaging other nations in developing and operating the global system of systems will promote the stewardship and solutions needed to address global change.

Today, the United States is not alone in either utilizing or investing in Earth observation capabilities. Earth observation products are increasingly being recognized around the world as essential for predicting global change and for developing policies addressing, mitigating, and adapting to global change. As noted above, Earth observations enable businesses and planners to better address the economic shocks arising from disruptions in water, food, energy, transportation, and logistics networks caused by global change. Similarly, defense and intelligence communities rely on Earth observation products to help them understand and predict the nature of the security threats arising from global change. Given the worldwide economic and global security concerns associated with global change, Earth observations clearly have a role in foreign policy formulation.

U.S. investments in Earth observation capabilities and Earth observation data policies are characteristic of U.S. scientific and political leadership in the international arena, particularly in participation in activities such as the IPCC and the integration of Earth observations into foreign aid and disaster relief activities.

Application: International Scientific Fora

Earth observations have played a key role in providing higher confidence in the scientific findings of the IPCC. Established by the WMO and the UN Environment Program, the IPCC assesses the scientific, technical, and socioeconomic information needed to understand climate change, its potential impacts, and options for adaptation and mitigation. Since 1990, IPCC assessment reports have contributed to international frameworks and dialogue related to climate change. These assessments have contributed to the UN Framework Convention on Climate Change, adopted in 1992 and entered into force in 1994, and the Kyoto Protocol, adopted in 1997. In its Fourth Assessment Report, the IPCC highlighted the increased role of Earth observations in providing a stronger quantitative basis for estimating likelihoods for future climate change. The IPCC concluded that scientists now have much higher confidence in their findings due to advances in climate modeling, as well as the collection and analysis of additional data. The United States is widely recognized for its contributions to the IPCC assessments stemming from not only its investments in Earth observation research capabilities but also making this data openly available to scientists worldwide.

Application: Development Aid, Disaster Relief and Foreign Policy

Tracking migrant populations, providing data to disaster rescue teams, monitoring widespread crop failure and drought, and planning the future growth of megacities are all examples of foreign policy decisionmaking opportunities reliant on Earth observations. Decisionmaking in all of these cases is reliant on Earth observation technologies and systems and forms an ever-growing portion of the broader foreign policy portfolio of the United States. Some examples, such as the use of Google Earth to highlight the devastation in Darfur, have even allowed the disintermediation of foreign policy and the interest aggregation of single individuals around the globe by bringing the scope of the conflict into every household. Other applications are more obvious; for instance, rescue crews providing relief after the December 2004 Indonesian Tsunami used Earth observations to locate villages that had been cut off, but not wiped out, and were, as a consequence, in dire need of food and clean water. In 1999, the European Space Agency (ESA) and Centre Nation d'Études Spatiales (CNES) initiated the International Charter for Space and Major Disasters (http://www.disasterscharter.org/), an international agreement among space agencies to support, with space-based data and information, relief efforts in the event of emergencies caused by major disasters. Space agencies in Canada, India, Argentina, Japan, and the United States (NOAA and USGS) have since joined (although the United States has not yet become a full-fledged member). To date, the charter has provided support in over 140 disasters around the world.

The U.S. government considers data from civil Earth observation systems to be a global public good and consequently promotes an open data policy. This open data policy, along with the level of U.S. investment in Earth observations and its participation in international fora, has characterized U.S. leadership in multilateral Earth observation discussions.

Background: GEO—A Multilateral Approach

In July 2003, the United States, under the leadership of Secretary of State Colin Powell, Secretary of Commerce Donald Evans, Secretary of Energy Samuel Bodman, and NOAA administrator Conrad Lautenbacher, launched a worldwide effort to develop a Global Earth Observation System of Systems (GEOSS). Today, 72 nations, the European Commission, and 52 other participating organizations are members of the Group on Earth Observations (GEO), the implementing body for GEOSS.

The U.S.-led creation of GEO has been a tremendous scientific, environmental, and foreign policy achievement. It has engaged—at the ministerial level—governments who have agreed on the value of Earth observations to obtain concrete societal benefits. GEO members have identified nine broad societal benefit areas that are intrinsically reliant on Earth observation products: disasters, health, energy, climate, agriculture, ecosystems, biodiversity, water, and weather. To date, GEO has focused on coordinating the data from applications in these nine areas, providing easier and more open data access and fostering use of this data for broader development of science, applications, and Earth observation capacity building. This coordination includes promoting architecture and software interoperability allowing the development of standard interoperable formats for collecting, processing, storing, and disseminating the full range of Earth observation data and thematic products. GEO is also developing a Web portal to provide a one-stop source for aggregating and accessing all the data from all the systems operated by its members. GEO is focusing on the transition from technology demonstration systems to a sustainable system for maintaining long-term data acquisition and continuity. GEO is also looking at current and new schemes for funding, including public-private partnerships.

Within GEO, the United States has led a growing consensus on making data freely available at low cost, prompting other nations to open up previously unavailable data sets. There was a broad recognition in the November 2007 Cape Town Declaration, agreed upon at the GEO Ministerial Summit in Cape Town, South Africa, that the success of GEOSS will depend on a commitment by all GEO partners to work together to ensure timely, global, and open access to data and products. Individual members made commitments in this direction. For example, Brazil and China offered their China-Brazil Earth Resources Satellite (CBERS) data free of charge to all African nations, and Brazil is putting a ground station in South Africa to facilitate distribution of that data. The Cape Town Declaration stated that GEO supported "the establishment of a process with the objective to reach a consensus on the implementation of the Data Sharing Principles for GEOSS to be presented to the next GEO Ministerial Summit," which is expected to occur before the end of 2010.

Background: GEO—A Multilateral Future?

International cooperation is essential for the creation of GEOSS. The realization of GEOSS will be achieved through the timely combination of observations from around the world and a number of relatively small geographical regions. This will require the contributions of many nations, since no single country has the resources to build the global system of systems needed to address global change. It will also require an understanding of what gaps exist, where capabilities overlap, and where strategic redundancies are required. And it will require a multilateral commitment to leverage the capabilities of all nations that have or plan to have Earth observation programs.

Other nations are making commitments to and large investments in Earth observations, which can complement U.S. Earth observation capabilities and contribute synergistically to a global system of systems. There is strong European political support for monitoring and measuring global change, and the European Global Monitoring of Environment and Security (GMES) Initiative has gathered substantial momentum in the European Union and with European space agencies, as well as wide-ranging participation of European industry. The Japanese Space Agency (JAXA) launched the Advanced Land Observing Satellite (ALOS) in 2006 and is developing Global Change Observation Mission (GCOM-Water) and planning a GCOM-Climate mission. China is making significant investments in Earth observations, with plans to launch several satellites for Earth resources, meteorology, and oceanography in the next decade. India is developing a suite of instruments and is planning a joint mission with France (Megha-Tropique) focusing on the water cycle in the intertropical region. A list of current and planned Earth observation satellites is included in appendix B.

With Europe, China, Japan, India, and other nations making very large investments in Earth observation capabilities, there is tremendous potential for synergistic cooperation and novel ways to leverage new capabilities. For example, the Europeans are now making decisions about the GMES program that could result in the (intentional or unintentional) development of a capability that already exists in other programs elsewhere. In the past, some nations have elected to duplicate capabilities existing elsewhere because they want their own "national" satellite or because they want a better capability. Another factor driving these decisions about redundancy is data availability. For example, in response to data sharing difficulties and the national security applications of some data, the United States wants to maintain a core capability for land imaging that would be synergistically supplemented with data from other non-U.S sources. Nonetheless, even today, some satellite capabilities are needlessly duplicated as a result of lack of coordination.

The Committee on Earth Observation Satellites (CEOS) grew from an initiative launched at the 1982 G-7 Economic Summit to coordinate space-borne Earth observation missions. The main goal of CEOS is to ensure that critical questions relating to Earth observations and global change are addressed while avoiding the unnecessary duplication of satellite missions related to global change. CEOS has 28 members that have satellite programs and an additional 20 associate members, including UN agencies and organizations that have or plan to have relevant ground facilities, support programs, or satellites. CEOS is a GEO Participating Organization: the GEO Secretariat and many GEO members look to CEOS to help assemble the space segment of GEOSS. The CEOS Strategic Implementation Team is developing several virtual satellite constellations involving data sharing. While CEOS leads or contributes to a number of GEO tasks, including the "virtual constellation task," the relationship between GEO and CEOS with respect to the development of GEOSS is still evolving and could be strengthened.

Furthermore, much more needs to be done in order to implement GEOSS. Jason and NPOESS are examples of successful international engagement. There are further opportunities for the United States to be proactive in seeking partnerships on cooperative missions and developing interoperable systems. With the exception of ocean monitoring, the United States has not built the cooperative relationships to transition new sensors and systems (beyond what are essentially technology demonstration missions) to long-term data acquisition and continuity.

Finally, it is widely acknowledged that strong, senior U.S. leadership in the GEO process has been critical to achieving the gains it has made. NOAA administrator Lautenbacher has played a key role in the creation and continuous U.S. leadership in GEO. It is critically important that the United States continue to represent itself in the GEO arena at such a senior level in order to keep ministerial level officials in other nations engaged and supportive of the creation of GEOSS. The priority that the United States places on GEO as evidenced by the leadership it provides will have a significant impact on the future of GEO.

Background: Export Control and Earth Observations

Export control regulations are a fundamental disincentive and significant structural impediment to U.S. participation in international systems, to foreign cooperation with the United States, and to the development of GEOSS. Since Earth observations can involve airborne, oceanic, and ground observation and are not limited only to space systems, the International Traffic in Arms Regulations (ITAR) legislation does not automatically come into play in every discussion of U.S. cooperation in GEOSS. However, ITAR can make it difficult to even initiate discussions on potential collaboration in many fruitful and obvious areas. ITAR has created real and perceived obstacles to engagement and cooperation. Although ITAR was intended to cover critical, highly sensitive military technologies, in practice the regulations are applied to a much, much wider array of other technologies. In addition, as individuals in the approval process are criminally liable equally for real and perceived mistakes, decisionmakers have a strong incentive to be excessively cautious. Furthermore, ITAR is now being applied to data from space systems, not just the space systems themselves. This has led to a situation where ITAR has forced the international community to develop their own independent capabilities (for example, radar ocean altimetry and Lidar/IMU). As a result, the international community now leads in several technologies, and U.S. firms are losing access to global markets and in some cases have lost the ability to produce such technologies altogether.

In January 2008, President Bush signed a package of U.S. Export Control Reform Directives to make the export licensing process more efficient and transparent, but the directives do very little to address the fundamental obstacles created by current export control regulations. Under the directives, additional funding will be provided for the State Department–led review of ITAR export license requests, and a 60-day deadline was set for decisions on applications. The directives are also intended to provide more transparency in the Department of Commerce's dual-use export control system. While the Commerce Department indicated it would be developing a regular process for systematic review of the list of controlled dual-use items, the directives themselves did not remove any items from either the dual-use or ITAR control lists.

Chinese factory worker assembling precision parts for electronic components.

Photo of the eyewall of Hurricane Katrina over the Gulf of Mexico, as seen from a NOAA P-3 hurricane hunter aircraft.

2 WHERE ARE EARTH OBSERVATIONS TODAY?

Who in the U.S. Government Is Responsible for Earth Observations?

Despite the profound interlocking uses and consequences of Earth observations and global change, no one department, agency, or person is responsible for setting a national vision for Earth observations; developing and leading the planning process; implementing the Earth observation architecture; developing and advocating budgets; or planning for the next generation Earth observation system. Responsibilities for Earth observation system activities are spread around several federal agencies. Furthermore, Earth observations are not a priority mission for any agency at the cabinet level. NASA is an independent, noncabinet level agency. NOAA resides in the Department of Commerce, while the USGS is within the Department of the Interior, and as such, all compete for funding with many other departmental responsibilities and priorities.

The United States has established a coordinating body called the U.S. Group on Earth Observations (USGEO) involving 17 federal agencies. Established in April 2005, USGEO is a standing subcommittee reporting to the National Science and Technology Council's Committee on Environment and Natural Resources in the Executive Office of the President. USGEO is currently developing a U.S. Earth observations policy to establish processes through which decisions can be made on the state of current and future U.S. Earth observations. It is hoped that these processes will address topics such as the scope of Earth observations; roles and responsibilities of federal agencies; decisionmaking; architecture design and implementation; interoperability; transition from shorter- to longer-term research; data management, archiving, and availability; and coordination of international partnerships. USGEO is developing an assessment and planning process to define national Earth observation needs and associated budgets through assessing current Earth observation capabilities, identifying gaps and overlaps, establishing priorities, identifying responsibilities of federal agencies, and coordinating budget planning with federal agencies and the U.S. Office of Management and Budget (OMB). USGEO is also working to become the focal point for U.S. Earth observation activities at the federal, state, and local levels and with private-sector, academic, and international partners.

Though these efforts are a laudable attempt to address the dispersion of responsibility for Earth observations in the U.S. government, there are questions about whether they are enough to address the looming gaps in Earth observation coverage, current and future investments needed for Earth observation systems, and the lack of planning for future Earth observation systems. USGEO is a coordinating body that has three cochairs, has no budgetary authority for Earth observations and cannot compel federal agencies to take any actions with respect to budgeting for Earth observation capabilities or implementing a national Earth observation plan. USGEO also cannot compel governmental agencies to work together to more effectively transition research capabilities to a system that acquires data consistently over the long term. This arrangement limits the support for a robust, holistic national Earth observation program to meet the nation's needs and a national budget sufficient to implement it.

What Challenges Face Earth Observations?

Challenge: Science and Applications

The majority of today's Earth observation systems are focused on science and examining how and why the Earth is changing; how the Earth responds to these changes; the impact of these changes; and how the Earth system will change in the future. In the *Earth Science and Applications from Space* decadal survey,[1] the Committee on Earth Science and Applications from Space wrote that there is a need for science missions that are focused on longer-term science questions. Continuous information is needed not only on what changes are taking place at this instant, but also on the trend and pace of change. Long-term sustained measurements are needed to support both short-term predictions, through consistent provision of data, and science, by providing records of Earth processes over many years.

However, U.S. Earth observation systems are not collectively well suited to address the impacts of climate change, its variability and intensity, water use, or many other changes that occur over long time scales. If our Earth observation systems were better suited for long-term data acquisition and continuity, our existing observation capabilities could provide real value for research applications by monitoring land use and cover change, agriculture delineation, fires, vegetation phenology and properties, forests, water supply and quality, and atmospheric aerosols. Monitoring developments in these areas would provide valuable information needed for making critical decisions. However, the national security community has become increasingly aware that current climate modeling lacks the geographical granularity to predict the long-term effects of global change at subregional and national levels in all regions of interest with sufficient accuracy to meet the detailed long-term planning needs of national security decisionmakers.

Challenge: Research to Operations Transition

The United States has led the world in realizing the global interconnectedness of people and the environment as well as leading the development of a new scientific field called global integrated Earth system science. The science community has already determined the set of key observables that must be measured in order to monitor the Earth system effectively. NASA has developed a number of measurement sensors and techniques—64 sensors on 18 spacecraft in 2005—and has demonstrated their vast capacity. Today, NASA is continuing to conduct research and development of sensors and monitoring techniques. Building on these proven capabilities, the United States now operates a highly effective national weather prediction system that has saved countless lives and billions of dollars.

The United States made the decision to manage operational weather satellites and has plans in place for satellite coverage into the mid 2020s. NOAA currently procures and operates the Geostationary Operational Environmental Satellites (GOES) and the Polar-Orbiting Environmental Satellites (POES) systems. The National Polar-orbiting Operational Environmental Satellite System (NPOESS) is being developed to ensure continuity of U.S. polar satellite observations after the

1. Committee on Earth Science and Applications from Space, National Research Council, *Earth Science and Applications from Space: Urgent Needs and Opportunities to Serve the Nation* (Washington, DC: National Academies Press, 2005).

final launches of NOAA's POES system and the Department of Defense's Defense Meteorological Satellite Program.

Land imaging in the United States has been carried out with the Landsat series of missions. The Landsat satellites have always been primarily a program to develop sensors and Earth observation technologies, rather than research involving long-term data acquisition and continuity. Yet, in addition to scientists, many private-sector users are now relying on these technology testbeds to produce data used in an array of Earth observation products. For this reason, the U.S. government decided it could not continue to provide land imaging on an ad hoc basis, which led to the August 2007 decision to establish a National Land Imaging Program to assume funding and management control of U.S. land imaging capabilities and applications with the objective of addressing, over the next year, both the development of new capabilities and the long-term data acquisition and continuity needed to support current and future Earth observation applications.

There are currently no U.S. Earth observation missions that deliver climate measurements as anything but an ancillary side effect of technology development activities. However, the future NPOESS will contribute to the continued measurement of more than half of the essential climate variables. In Europe, the European Organisation for the Exploitation of Meteorological Satellites (EUMETSAT) was created in 1986 through an international convention agreed to by 20 European member states (as of now) to deliver weather and climate-related satellite data, images, and products on a continuous basis. EUMETSAT supplies this data to the national meteorological services of the organization's 20 members and 10 cooperating states and other users worldwide. While the data and images from geostationary (Meteosat) and polar-orbiting (MetOp-A) satellites are used for weather forecasting, they also provide long-term data acquisition (and some measure of continuity) to scientists for climate change research. The United States, France, and EUMETSAT have been cooperating on the Jason satellites to conduct more in-depth research on the Earth's oceans, which is leading to the provision of a future capability (Jason-3) in 2013 that may extend data acquisition and continuity in this area.

NOAA and NASA are working toward establishing an interagency transition office to address the connection between research and development activities that demonstrate new observation capabilities and the longer-term acquisition and continuity of data. The agencies are focusing their efforts on near-term transition opportunities and hope to demonstrate such a transition in the near future using Jason-3, with the transition of Ocean Surface Vector Winds observations as a possible future transition initiative.

In the strategic plan for the U.S. Integrated Earth Observation System, USGEO identified nine domestic societal benefit areas[2] (which are the same as the nine societal benefit areas identified by GEO, except that USGEO includes oceans and does not include biodiversity). Other than missions addressing weather, there has never been a U.S. Earth observation mission focused primarily on a societal benefit area. The research and operational communities are examining the feasibility of accomplishing short- and long-term technological, scientific, and operational objectives on the same platforms through better coordination and expanded use of observations. Episodic events could require broader geographical or chronological coverage than technology demonstration missions

2. The nine areas are: (1) improve weather forecasting; (2) reduce loss of life and property from disasters; (3) protect and monitor our ocean resources; (4) understand, assess, predict, mitigate, and adapt to climate variability and change; (5) support sustainable agriculture and forestry and combat land degradation; (6) understand the effect of environmental factors on human health and well-being; (7) develop the capacity to make ecological forecasts; (8) protect and monitor water resources; and (9) monitor and manage energy resources.

can deliver; however, in some cases, both technology and applications development use similar information on the planet and how it is changing. Do climate observations require dedicated Earth observation resources? Developing and operating dual systems for both long-term data acquisition and continuity and technology demonstration purposes may not be a luxury that the United States or any other nation can afford.

In sum, the aggressive U.S. research and development program has developed a number of proven sensors and ways of measuring data that have yielded new scientific understanding and short-term forecasting improvements. However, due to structural and budgetary factors, these gains in obtaining new data have not resulted in the creation of the continuous, complete, and comprehensive data sets needed to improve our understanding of the longer-term—and perhaps much more important—climate changes that lie at the core of many current policy debates. There is no U.S. government commitment to continuous acquisition of data over the long term that will be critical for a better understanding of why, how, and how fast the Earth is changing. Even more troublingly, there is no effective U.S. government planning for a future comprehensive, coordinated, and sustainable national Earth observation system to gather data over long time scales.

Challenge: Earth Observation Requirements

Today, the set of commonly recognized observables consists of 55 weather observables identified by the NPOESS Senior Users Advisory Group; 26 essential climate variables identified by the Global Climate Observing System; and 5 solid Earth strategies identified by the NASA Solid Earth Science Working Group. Earth observation users from other sectors, such as industry users of Earth observation data, have not yet had the opportunity to provide input to a national set of Earth observation requirements, which may inhibit private-sector use and support of national Earth observation capabilities.

Challenge: Program Management

There are also many questions and concerns surrounding the agency-level management of Earth observation systems, particularly space-based capabilities. NASA has traditionally procured space-based Earth observation systems, such as the weather and land imaging satellites, even when other agencies have been responsible for their operation. Under the new National Land Imaging Program, NASA will continue to procure Landsat satellites, although USGS for the first time is acquiring the ground segment under the Landsat Data Continuity Mission (LDCM) program. A tri-agency approach was adopted for the NPOESS system. NOAA, NASA, and the Department of Defense are jointly responsible for developing, acquiring, managing, and operating NPOESS. The involvement of three federal agency bureaucracies in the acquisition phase has proven to be challenging. No clear answers have emerged suggesting the most effective and efficient division of roles and responsibilities for Earth observations programs during the research, transition, and operational phases. This planning shortfall becomes a critical issue with the potential implementation of the so-called cap and trade agreements for carbon emission management. Cap and trade agreements will both need strong verification mechanisms and as an understanding of how royalties from cap and trade programs will be managed. The management experiences associated with NPOESS and other programs will be important lessons when making decisions on the management of the next generation Earth observation system.

Challenge: Platform Mix

The United States currently uses a mixture of Earth observation systems including remote sensing satellites and in-situ capabilities. Traditionally, the United States has relied mainly on large, space-based platforms that carry a suite of sensors and less on small, single-purpose satellites like Jason-1 (weighing roughly 500 kilograms or approximately 1,100 pounds) that can be developed more quickly and provide complementary capabilities to larger satellites. New capabilities are also emerging such as Unpiloted Aerial Vehicles (UAVs) and buoys that could provide additional observation opportunities.

How Will We Plan and Pay for Earth Observations?

Ironically, at a time when U.S. public-sector investments over the past two decades are enabling the observations needed to identify and understand ongoing global changes, the United States is failing to invest enough to maintain the current capability over the next two decades. Today, the United States has the Earth observation satellite capacity to monitor many essential climate variables, but gaps in coverage will appear in the next decade and a half, and there are no current plans for capabilities beyond 2025.

Challenge: Budget Shortfalls

Last year, the National Academy of Sciences urged the U.S. government to increase its investments in Earth observation satellites. Its decadal survey *Earth Science and Applications from Space*[3] warned that the existing constellation of Earth observation satellites will degrade significantly over the next decade due to insufficient program funding. The report noted that the replacement sensors to be flown on NPOESS are generally less capable than the current (space-based) Earth Observation System (EOS) counterparts they are intended to replace, and it recommended that NOAA restore several key climate, environmental, and weather observation capabilities that had been removed during the 2006 NPOESS restructuring. The National Research Council also recommended funding the 17 highest-priority Earth observation satellite missions between 2010 and 2020, or the United States will risk losing up to 75 percent of its satellite monitoring capability.

There are currently 14 satellite missions on orbit comprising NASA's EOS. The NASA Terra, Aqua, and Aura spacecraft are large observatories that carry suites of precision sensors to acquire terrestrial, oceanic, and atmospheric data. These three satellites, along with CALIPSO, CloudSat, and PARASOL developed by the Centre National d'Études Spatiales (CNES), are in specific orbits to provide near-simultaneous measurements. NASA also has seven missions under development that will be launched between 2008 and 2013. However, in 2011, it is estimated that only five to nine NASA missions will be operating, potentially constituting a reduction of over 50 percent. To begin to address this shortfall, the president's fiscal year (FY) 2009 budget request for NASA included $103 million to start two new Earth science missions identified as priorities in the National Academy's survey. The budget request also supports three additional new starts (mission to be determined) over the next six years. However, this represents only 5 of the 17 missions that the survey recommended.

---

3. Committee on Earth Science and Applications from Space, *Earth Science and Applications from Space.*

Climate sensors were shifted to NPOESS, which is expected to provide capabilities through 2026, from NASA's EOS during the Clinton administration. However, these climate sensors were removed during the restructuring of NPOESS in June 2006 due to budget problems. In April 2007, the Ozone Mapping and Profiler Suite (OMPS) Limb was restored on the NPOESS Preparatory Project. During the remainder of 2007, NASA, NOAA, and the Office of Science and Technology Policy (OSTP) discussed options for flying the other sensors either on NPOESS or another platform because of the concern of returning risk to that program; however, they took no action to preclude the possibility of still flying the sensors on NPOESS. To respond partially to the *Earth Science and Applications from Space* report's recommendations to restore the sensors removed from NPOESS, the president's 2009 budget request for NOAA included $74 million to restore two more climate sensors. The already-built Clouds and the Earth's Radiance Energy System (CERES) sensor will be put on the NPOESS Preparatory Project. NOAA would also start work on a Total Solar Irradiance Sensor and look for a suitable spacecraft for it.

In the United States, Landsat satellites have never been considered a fully operational capability, and no single U.S. government agency has had the responsibility for meeting U.S. needs for operational moderate-resolution ground imaging. Over the years, many attempts were made to commercialize the provision of moderate-resolution ground imaging data, but a viable commercial option never emerged. Consequently, the United States has been unable to adequately address the expected gap in U.S. moderate-resolution land imaging data. Technical problems with the current Landsat 5 and 7 satellites are expected to result in their unavailability prior to the 2011 launch of the LDCM. In addition, there currently is no successor mission to LDCM nor a replacement satellite should LDCM fail at launch or early in its operational life. The new National Land Imaging Program provides a focal point in the U.S. government for understanding land imaging requirements and planning and budgeting for missions to meet these requirements. USGS has begun working within the Department of the Interior to begin to migrate the current Land Remote Sensing Program into the National Land Imaging Program. However, the Interior Department did not receive additional funding last year to implement these new responsibilities, and only $2 million was requested for this by the administration for FY 2009. USGS is currently coordinating and promoting the uses of land imaging data within the Department of the Interior.

There is strong concern about the limited amount of resources being applied to the data management side of Earth observations. The Earth observation system is an end-to-end system encompassing the information chain from data to knowledge. Currently, there are not sufficient computing capabilities to make full use of the observations that are being collected for anything beyond the simplest of technology demonstration purposes. The United States is building the platforms and the platforms are consuming a very large portion of the available resources, but the remainder of the Earth observation value chain, including data management, is not getting sufficient funding. It is estimated that approximately $100 million to $200 million annually is needed to configure the decision support process and tools to enable the United States to obtain adequate benefit from Earth observations and serve the policymakers and managers who need the data. A similar level of funding is needed to improve and enhance weather, climate, and hazards forecast models to optimize the benefits from Earth observations.

Another issue that needs to be considered in the planning and budgeting for the next generation space-based Earth observation system is which launch vehicles will be used. With the Delta 2 no longer available, there is a concern that more funding will be required for launchers, with consequently less funding available for the platforms and instruments.

In an increasingly constrained "discretionary" fiscal environment, NASA, NOAA, and USGS budgets for Earth observations, which began to decline in FY 2006, are competing with other national priorities, and existing budgets have not been sufficient to either maintain current capabilities, meet clearly identified current needs, or provide the Earth observation products needed for foreign policy, economic, or security applications. For example, the president's budget requests for NASA since the Vision for Space Exploration (VSE) was announced in January 2004 have been approximately $2 billion below the administration's five-year budget projection in FY 2004. The Earth science community estimates that the overall Earth observation budget is underfunded on the order of $2.5 billion annually. Moreover, funding has become unbalanced among dedicated defense and intelligence capabilities on the one hand and purely civil space activities on the other, even though civil space activities such as Earth observations provide very significant value to other sectors.

Challenge: Planning for Next Generation Earth Observation System

The National Academy's *Earth Science and Applications from Space* decadal survey did not address next generation Earth observation system requirements in the post-NPOESS era (2025 and later), and there is currently no U.S. planning underway for this period. Historically, EOS and NPOESS were initiated roughly every 10 years (1986 and 1994, respectively). Based on experience, the development and deployment of space-based Earth observation systems takes an average period of 15 years from initiation of program plans and funding until launch. This strongly suggests that planning for space-based Earth observation capabilities beyond NPOESS and LDCM should already have been initiated if we are to avoid another major degradation of capability as NPOESS and the Landsat satellites age. Furthermore, there is no planning underway for a future, comprehensive, coordinated, and sustainable Earth observation system to continuously acquire data over the long term.

The traditional U.S. government budget cycle creates problems with respect to long-term planning of Earth observation systems; additionally, future systems cannot be based on agency priorities alone but need to be planned to meet national requirements. These broader national requirements must be derived from input from federal agencies, state governments, academia, and a wide array of industry and private-sector users. There has been no significant community input to the OSTP process to address future Earth observing systems as yet.

How Will the Public and Private Sectors Work Together?

Currently, the private sector is both a supplier of Earth observation systems to the government and a user of those same space assets. There are many good public-private partnerships involving the use and dissemination of Earth observation data from government-purchased weather and land imaging satellites. However, past U.S. government attempts to commercialize space-based Earth observation capabilities have not been successful. And there are a large number of Earth observations (space and terrestrially based) that are being made by private-sector actors that are not being effectively used by the government.

As mentioned previously, U.S. attempts to commercialize moderate-resolution land imaging observations have not been successful. The *Land Remote-Sensing Commercialization Act of 1984* established a policy to commercialize those remote-sensing space systems that properly lend themselves to private-sector operation and avoid competition between the government and commercial operations. Congress concluded that the private sector, in particular the "value-added industry," would be best suited to commercially develop land remote sensing. Recognizing that the private sector alone could not develop a total land remote-sensing system because of the high risk and large initial capital expenditure involved, Congress called for cooperation between the federal government and private industry to assure both data continuity and U.S. leadership to achieve a phased transition to a fully commercial system. However, it said cooperation should be structured to involve the minimum practicable amount of support and regulation by the federal government and the maximum practicable amount of competition by the private sector, while assuring continuous availability to the federal government of land remote-sensing data. In 1992, Congress concluded that full commercialization of the Landsat program could not be achieved in the foreseeable future and repealed the 1984 legislation. This change in U.S. policy recognized the reality that the costs of building, launching, and operating the satellites, processing and registering the data, and analyzing and providing the data as value-added services effectively prevented the United States from successfully commercializing Landsat. In both the 1984 and 1992 Acts, Congress indicated there was no compelling reason to commercialize meteorological satellites. In 1994, the U.S. government changed its policy on high-resolution land imaging capabilities and agreed to provide a license to the private sector to build, operate, and commercialize Earth observation optical systems up to one-meter resolution. This policy was successful, with two U.S. companies now dominating this market.

Europe has been much more successful in commercializing land imaging capabilities, primarily because of its unique cost-sharing arrangements that involve significant government support. The French company Spot Image operates Earth observation satellites and sells a wide range of imagery and geo-information products and services. Under a public-private partnership, CNES, the French space agency, finances the satellites, and Spot Image provides funding for the ground segment and operations. The government purchases the data it needs, which helps cover a marginal (less than 10 percent) portion of the operations costs. Through this arrangement, the government saves between $25 million and $35 million per year. Spot Image has partnered with EADS Astrium to initiate the first dual-use and public-private partnership involving the high-resolution stereoscopy (HRS) instrument on Spot 5. The government provides 46 percent of the funding, with industry picking up the remaining 54 percent. The HRS mission itself is carried out jointly. While Spot Image generally has exclusive distribution rights, the French Ministry of Defense retains some specific distribution rights to allow for various data exchange among allied forces. European industry and government are partnering on AstroTerra, a Spot 5 follow-on, which will reduce government's costs by $22 million per year. The public-private partnership makes it easier for the government to provide funding since the investment is spread over many years and keeps European industry competitive in the international market.

There are many other mechanisms for effective collaboration among contributing segments of the community and emerging users, such as the recent NOAA partnership with Shell to place sensors on their offshore oil platforms, which will augment oceanic data sets. The role that the private sector can play in putting together GEOSS is being examined more closely. Currently, GEO is exploring both traditional and novel mechanisms for funding and using Earth observing systems, including public-private partnerships. GEO is exploring a partnership with Iridium to put sensors on Iridium's NEXT generation LEO satellite constellation. GEO is also working with NOAA, EUMETSAT, and other meteorological organizations to develop GEONETCast to disseminate space-based, airborne and in-situ data and products to users worldwide.

There is a growing recognition that as private-sector dependence on Earth observations grows, the level of support for such capabilities also needs to increase to sustain this continued growth; in addition, the private sector will need to increase its support of these capabilities to ensure the continued and future availability of necessary information in a complete and timely fashion independent of the uncertainties of global public investment. Once the market matures and the economic value of Earth observations is more clearly established, there will be more opportunities for public-private partnerships and private-sector initiatives. Other emergent opportunities, such as the possibility of earthquake prediction, may be fertile areas for involvement of the private sector at an early stage of the development process. More effective integration of the entire Earth observation value chain will provide a broader base of support and capital investment for a more comprehensive global Earth observation capability meeting both private- and public-sector needs.

Landsat 7 undergoing clean room inspection prior to launch.

In addition to the many technological challenges presented by global change, a wide array of regulatory and policy issues remain unresolved.

3 WHAT HAPPENS NEXT IN EARTH OBSERVATIONS?

There are many policy, programmatic, and budgetary issues related to near- and long-term Earth observations that need to be urgently addressed if the Earth observation capabilities that governments, academia, industry, and others have come to rely are to be available in the future. These capabilities will be needed more than ever as the incidence and pace of global changes continue to increase in order to maintain our ability to understand and identify ways to predict, prevent, and mitigate the impacts of these changes. Importantly, without these Earth observation capabilities, we run the very real risk of misdirecting our resources toward solving the wrong problems.

Several national and international strategies have been identified to address the issues facing Earth observations. These recommendations are also included in the Executive Summary.

Value of Earth Observations and U.S. Challenges and Opportunities

Recommendation 1. The United States should make a commitment to long-term, continuous data acquisition for all essential observations necessary to provide improved monitoring and prediction capabilities in order to sustain monitoring and evolve our understanding of the Earth system and how it is changing.

Recommendation 2. Building on the National Academy's decadal survey *Earth Science and Applications from Space*, the United States should develop an overall plan for an integrated, comprehensive and sustained Earth observation system that (1) describes how these measurements can not only be acquired, archived, and distributed but also integrated as appropriate with Earth science models and decision support tools and (2) maps the goals and requirements for this system to and from the nine societal benefit areas identified by the U.S. Integrated Earth Observation System. Shortfalls in the current plans should be addressed, and a vision for future generation Earth observation systems should be provided. Users from all sectors—public (federal, state and local governments), academia, and industry—should have the opportunity to provide input and participate in the definition and planning process.

Recommendation 3. The U.S. government should increase funding for Earth observations by doubling the budget from approximately $2.5 billion to $5 billion annually. This would enable expanding both space-based and in-situ observational capabilities to fully implement the National Academy's *Earth Science and Applications from Space* decadal survey recommendations and increase supercomputing, decision support tools, and modeling capabilities. This funding level should be reassessed following the development of an overall architecture for Earth observations in order to adequately fund an integrated, comprehensive, and sustained Earth observation system for weather, climate, hazards, Earth science, and resource management—consistent with the goals of GEOSS.

Recommendation 4. The United States should establish a governance structure under the supervision of a cabinet-level position that provides leadership both within the United States for Earth observations and for coordination and cooperation with other nations to promote and facilitate the planning, funding, and implementation of GEOSS. A formalized high-level interagency process should be established at the White House, led by a cabinet-level administration official responsible for U.S. Earth observation vision and goals, to establish agency roles and responsibilities for an integrated, comprehensive, and sustained Earth observation system and to develop the integrated budget to implement it. The president should provide an annual report on the state of the environment based on Earth observation measurements and on the status of and issues related to the development and operation of the comprehensive Earth observation system.

International Challenges and Opportunities

Recommendation 5. The United States should leverage its investment in Earth observations as a foreign policy tool, not only to promote global stewardship and develop global solutions but also to enhance U.S. soft power capabilities and improve its international image. The U.S. government should formally join the International Charter for Space and Major Disasters.

Recommendation 6. The United States should optimize international partnerships for the development and operation of Earth observation capabilities that leverage global synergies to minimize gaps and unnecessary overlaps while providing strategic redundancies.

Recommendation 7. The U.S. government should continue to support the multilateral GEO at a senior level to help actively promote the development of GEOSS and maintain the engagement of other governments at the ministerial level. Through its support of GEO, the United States should motivate the international community to make the investments necessary to build a coordinated, worldwide Earth observation system of systems. It should also continue to provide leadership and support to CEOS to define the space segment of GEOSS.

Recommendation 8. The multilateral GEO, with the support of CEOS, should build on the progress it has already made to champion and lead the process of developing a truly comprehensive, coordinated, worldwide Earth observation system of systems for all Earth observation needs. In doing so, current and planned capabilities should be leveraged so that both gaps and unnecessary overlaps are eliminated [**this repeats a line from Recommendation 6**]. Building on the commitments already made by some GEO members, GEO should promote the adoption of an open data policy by all its members as a goal to be aspired to.

Recommendation 9. The State Department should continue to include the important role of Earth observations in general and the GEO in particular in responding to global change at the G-8 summit and in other high-level multilateral and bilateral fora to raise the visibility of Earth observations and global change at the highest levels of government.

Recommendation 10. The United States should revise its export control policies to promote dialogue among international governmental and industrial partners. The U.S. government should review the ITAR list to remove nonsensitive items related to Earth observations that present no threat to national security in order to promote dialogue among industrial partners, create a healthier climate for U.S. business, and achieve more international collaboration on Earth observations. Furthermore, ITAR should no longer inhibit the exchange of scientific data, in order to remove fundamental disincentives to the discussions that lead to cooperation.

Private-Sector Challenges and Opportunities

Recommendation 11. The private sector should be an active participant in the development of the architecture for an integrated, comprehensive, and sustained Earth observation system to ensure its requirements are generated and incorporated in a holistic way and to enable insight into ways that a mature private-sector capability can evolve to address potential gaps. Recognizing U.S. government challenges in space systems acquisition management, the private sector should be called on to bring ingenuity and innovation to solutions for future system configurations, technologies, and delivery that focus on maximizing performance while reducing risks, shortening development time, and minimizing budgets.

Recommendation 12. The U.S. government should actively seek innovative public-private partnerships for developing and operating Earth observation systems that will capitalize on or promote the emergence of private-sector capabilities. In addition to the traditional commercial Earth observation players, the increasingly broad range of industries that are ever more reliant on Earth observations should be engaged to ascertain how they cooperate in their roles as consumers or suppliers of Earth observation products and data throughout the entire Earth observation system in meeting the overall objectives of Earth observation and global change public policy.

Recommendation 13. To promote commercialization, the U.S. government should build on its history of providing no-cost or low-cost data (weather and most recently, Landsat) to promote the further development of private-sector products (for example, AccuWeather). If applications prove to be as profitable as they have been in other technology fields (such as the global positioning system for navigation), they could support the business case for private development and operation of more focused and tailored private-sector Earth observation offerings.

Recommendation 14. The U.S. government should engage the finance community to discuss and address how its policies and activities promote or fail to promote private capital and equity market investment in support of goods derived from Earth observation. The government should take steps to remove some of the long-standing impediments to creating public-private partnerships, such as the inability of the federal government to make long-term commitments to use assets provided by the private sector.

APPENDIX A
WORKING GROUP MEMBERS AND CONTRIBUTORS

SENIOR ASSOCIATE
Lyn Wigbels, *CSIS*

PROJECT MANAGER
G. Ryan Faith, *CSIS*

SENIOR FELLOW
Vincent Sabathier, *CSIS*

MEMBERS

Jose Achache, *Group on Earth Observations*
John Anderson, *Monsanto*
Bruce Betts, *The Planetary Society*
Ron Birk, *Northrop Grumman*
Don Blick, *Raytheon*
Bill Brennan, *National Oceanic and Atmospheric Administration*
John Cain, *Chevron*
Paul Carliner, *Carliner Associates*
Nancy Colleton, *Institute for Global Environmental Studies*
Paul Cooper, *Caris*
Frank Culbertson, *SAIC*
Antoine de Chassy, *Spot Image*
Michael Dykes, *Monsanto*
Ryan Engstrom, *George Washington University*
Louis Friedman, *The Planetary Society*
Sam Goward, *University of Maryland*
David Halpern, *National Aeronautics and Space Administration*
Maureen Heath, *Northrop Grumman*
Ravi Hichkad, *Northrup Grumman*
John Hillery, *CSIS*
Jim Hudson, *Monsanto*
Sarah Ladislaw, *CSIS*
James Lewis, *CSIS*
Johannes Loschnigg, *National Center for Atmospheric Research*
Molly Macauley, *Resources for the Future*
Jon Malay, *Lockheed Martin*
Stephen Moran, *Raytheon*
Linda Moodie, *National Oceanic and Atmospheric Administration*
Berrien Moore, *University of New Hampshire*
Timothy Newman, *U.S. Geological Survey*
Frank Nutter, *Reinsurance Association of America*
Richard Obermann, *U.S. House of Representatives*
Rick Ohlemacher, *Northrop Grumman*
Ian Pryke, *George Mason University*
James Plasker, *American Society for Photogrammetry and Remote Sensing*
Scott Rayder, *National Oceanic and Atmospheric Administration*
Daniel Reifsnyder, *U.S. State Department*
Barbara Ryan, *U.S. Geological Survey*
David Skole, *Michigan State University*
Brent Smith, *National Oceanic and Atmospheric Administration*
David Speiser, *SAIC*

Carla Sullivan, *National Oceanic and Atmospheric Administration*
John Townshend, *University of Maryland*
Laura Verduzco, *Chevron*
Darryn Waugh, *Johns Hopkins University*
Damon Wells, *Office of Science and Technology Policy*
Gene Whitney, *Office of Science and Technology Policy*
Eric Wieman, *Computer Sciences Corporation*
Ray Williamson, *Secure World Foundation*
Greg Withee, *National Oceanic and Atmospheric Administration*
Edmund Woollen, *SAIC*
Shira Yoffe, *U.S. State Department*
Jim Zimmerman, *International Astronautical Federation*

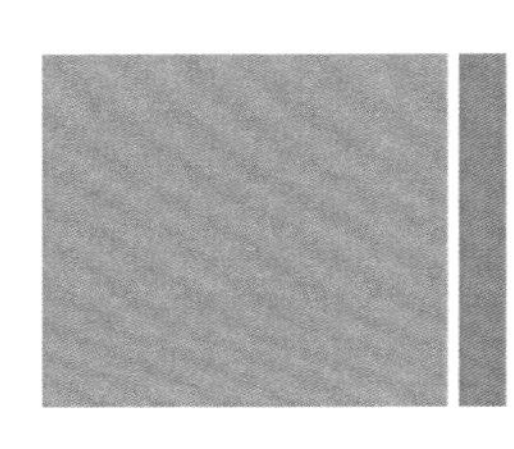

APPENDIX B
CURRENT AND PLANNED EARTH OBSERVATION SATELLITES

Launch Year	EO Satellite Mission (and sponsoring agency)
1967	Diademe 1&2 (CNES)
1975	STARLETTE (CNES)
1976	LAGEOS-1 (NASA)
1984	Landsat-5 (USGS)
	ERBS (NASA)
1990	SPOT-2 (CNES)
1991	METEOSAT-5 (EUMETSAT)
	NOAA-12 (NOAA)
	UARS (NASA)
1992	Topex-Poseidon (NASA / CNES)
	LAGEOS-2 (NASA / ASI)
1993	SCD-1 (INPE)
	STELLA (CNES)
1993	METEOSAT-6 (EUMETSAT)
1994	NOAA-14 (NOAA)
1995	GMS-5 (JAXA / JMA)
	ERS-2 (ESA)
	GOES-9 (NOAA)
	RADARSAT-1 (CSA)
1997	DMSP F-13 (NOAA)
	GOES-10 (NOAA)
	METEOSAT-7 (EUMETSAT)
1998	SPOT-4 (CNES)
	NOAA-15 (NOAA)
	SCD-2 (INPE)
1999	INSAT-2E (ISRO)
	Landsat-7 (USGS)
	IRS-P4 (ISRO)
	QuikSCAT (NASA)
	DMSP F-15 (NOAA)
	Terra (NASA)
	ACRIMSAT (NASA)
	KOMPSAT-1 (KARI)
2000	GOES-11 (NOAA)
	CHAMP (DLR)
	NOAA-16 (NOAA)
	NMP EO-1 (NASA)
	SAC-C (CONAE)
2001	Odin (SNSB)
	GOES-12 (NOAA)
	BIRD (DLR)
	TES (ISRO)
	Jason (NASA / CNES)
	TIMED (NASA)
	METEOR-3M N1
	(ROSHYDROMET / ROSKOSMOS)
2002	Envisat (ESA)
	GRACE (NASA)
	Aqua (NASA)
	SPOT-5 (CNES)
	NOAA-17 (NOAA)
	METEOSAT-8 (EUMETSAT)
	KALAPANA (ISRO)
	FedSat (CSIRO / CRCSS)
2003	ICESat (NASA)
	SORCE (NASA)
	INSAT-3A (ISRO)
	SCISAT-1 (CSA)
	UK-DMC (BNSC)
	RESOURCESAT-1 (ISRO)
	DMSP F-16 (NOAA)
	CBERS-2 (CAST / INPE)
2004	DEMETER (CNES)
	Aura (NASA)
	FY-2C (NRSCC)
	PARASOL (CNES)
	SICH-1M (NSAU / ROSKOSMOS)
2005	Meteor-M No1
	(ROSHYDROMET / ROSKOSMOS)
	Vulkan-Kompas-2 (ROSKOSMOS)
	MTSAT-1R (JMA)
	NOAA-N (NOAA)
	GOES-N (NOAA)
	CARTOSAT-1 (ISRO)
	CRYOSAT (ESA)
	Monitor-E (ROSKOSMOS)
	CALIPSO (NASA / CNES)
	CloudSat (NASA)
	DMSP F-17 (NOAA)
	TopSat (BNSC)
	METEOSAT-9 (EUMETSAT)
	Resurs DK (ROSKOSMOS)
	ALOS (JAXA)
	KOMPSAT-2 (KARI)
	TerraSAR-X (DLR)
	METEOR-3M N2
	(ROSHYDROMET / ROSKOSMOS)
	METOP-1 (EUMETSA T)
	Baumanets (ROSKOSMOS)
	CARTOSAT-2 (ISRO)
	RADARSAT-2 (CSA)
	SICH-2 (NSAU)
2006	BelKA (ROSKOSMOS)
	BISSAT (ASI)
	FY-3A (NRSCC)
	IGPM (ASI)
	Kanopus-V ulkan (ROSKOSMOS)
	LAGEOS-3 (NASA / ASI)
	TerraSAR-L (BNSC)
	MTSAT-2 (JMA)
	GOCE (ESA)
	INSAT-3D (ISRO)
	RESOURCESAT-2 (ISRO)
	RISAT-1 (ISRO)
	NMP EO-3 GIFTS (NASA)
	CBERS-2B (CAST / INPE)
	NPP (NOAA)
	Elektro-L
	(ROSHYDROMET / ROSKOSMOS)
	HJ-1A (CAST)
	HJ-1B (CAST)
	HY-1B (CAST)
	Jason-2 (NASA / CNES)
	COSMO - SkyMed (ASI)
	FY-2D (NRSCC)
	FY-3B (NRSCC)
2007	Meteor-M No2
	(ROSHYDROMET / ROSKOSMOS)
	OCEANSAT-2 (ISRO)
	RapidEye (DLR)
	SMOS (ESA)
	GOES-O (NOAA)
	HJ-1C (CAST)
	THEOS (GISTDA)
	OCO (NASA)
	SAOCOM 1A (CONAE)
	ADM-Aeolus (ESA)
	DMSP F-18 (NOAA)
	SAC-F (CONAE)
	Glory (NASA)
	SSR-1 (INPE)
2008	DSCOVR (NASA)
	ESA Future Missions (ESA)
	PICARD (CNES)
	GOSAT (JAXA)
	METEOSAT-10 (EUMETSAT)
	Pleiades 1 (CNES)
	SAC-D/Aquarius (CONAE / NASA)
	SAOCOM 1B (CONAE)
	GOES-P (NOAA)
	CBERS-3 (CAST / INPE)
	NOAA-N' (NOAA)
	FY-3C (NRSCC)
2009	GCOM-W (JAXA)
	Hyperspectral Mission (ASI)
	MEGHA-TROPIQUES (CNES / ISRO)
	Swarm (ESA)
	DMSP F-19 (NOAA)
	GPM Core (NASA)
	SAC-E/SABIA (CONAE)
	NPOESS-1 (NOAA)
	FY-2E (NRSCC)
	METOP-2 (EUMETSAT)
	Pleiades 2 (CNES)
2010	GCOM-C (JAXA)
	GPM Constellation (NASA)
	HYDROS (NASA)
	FY-3D (NRSCC)
2011	NPOESS-2 (NOAA)
	DMSP F-20 (NOAA)
	CBERS-4 (CAST / INPE)
	METEOSAT-11 (EUMETSA T)
2012	SAOCOM-2B (2) (CONAE)
	GOES-R (NOAA)
	SSR-2 (INPE)
	FY-3E (NRSCC)
2013	NPOESS-3 (NOAA)
	SAOCOM-2B (1) (CONAE)
2014	METOP-3 (EUMETSAT)
	FY-3F (NRSCC)
2015	NPOESS-4 (NOAA)
2016	FY-3G (NRSCC)
2018	NPOESS-5 (NOAA)
2019	NPOESS-6 (NOAA)

Source: Committee on Earth Observation Satellites.